Aquatic Earth

W9-AWG-003

Printed in Mexico

ISBN 978-0-15-362269-4
ISBN 0-15-362269-5

2 3 4 5 6 7 8 9 10   050   16 15 14 13 12 11 10 09 08

Harcourt
SCHOOL PUBLISHERS

Visit *The Learning Site!*
www.harcourtschool.com

The Water Planet

We often think of Earth as firm soil and rock upon which people and animals live. In fact, about 70 percent of Earth is covered with water. Water constantly flows above, across, and through Earth's crust and ecosystems in a process known as **water cycle**. Liquid water on Earth changes to water vapor and joins other gases in the atmosphere in a process known as **evaporation**. The water that has left eventually falls back to Earth. **Precipitation** is solid water (snow, sleet, hail) or liquid water (rain) that falls from the air to Earth.

Most precipitation falls into oceans, rivers, and lakes. Water that falls to the soil, enters the ground and is located within gaps in rocks is called **groundwater**. This is an important resource for humans and all forms of life on Earth.

The water cycle maintains a constant flow of water from Earth into the atmosphere and back again.

precipitation

evaporation

ocean

groundwater

Most of the water on Earth is salt water. Fresh water, water with a very low salt content, is only about 3 percent of the water on Earth. Three-fourths of that water is nearly impossible to use because it is frozen in ice caps and glaciers!

People cannot survive without fresh water. People use it for drinking, growing crops, cooking, and cleaning.

It is important to protect our water. Yet much of it is polluted or destroyed by oil, pesticides, and other toxins. In most places today, fresh water must go through water treatment plants before it is safe for people to use.

 SEQUENCE Describe the processes and places in a glass of water's travel from a river to your lunch table.

Water on Earth usually needs to be cleaned at a water treatment plant before it is ready for use by humans.

The Ocean Floor

You might think that the ocean gradually slopes down from the beach, and is smooth and flat like the bottom of a bowl. In fact, the floor of the ocean is as varied as the land on Earth. Oceanographers have used sonar, or sound wave technology, to map it. On the ocean floor, there are wide, flat plains, deep valleys, tall mountains, and even volcanoes.

The ocean floor can be divided into three major regions. The first region is the **continental shelf**, a gradually sloping part of the ocean floor. When you wade into the ocean from the beach, you are walking on the continental shelf. In most places it stretches for at least 30 km (19 miles) away from the shore.

Just past the continental shelf is the second region, the **continental slope**. Cliffs on the continental shelf begin a deep plunge to the ocean floor in this region. The average depth of the ocean is 3,720 m (12,200 ft).

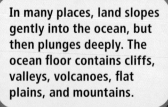

In many places, land slopes gently into the ocean, but then plunges deeply. The ocean floor contains cliffs, valleys, volcanoes, flat plains, and mountains.

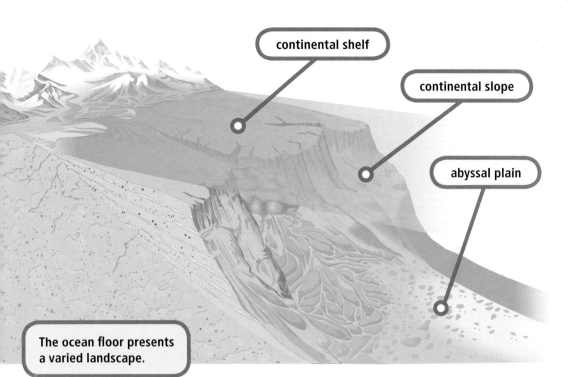

continental shelf

continental slope

abyssal plain

The ocean floor presents a varied landscape.

At the end of the continental slope, the floor of the ocean flattens out. This flat area of the ocean floor is called the **abyssal plain**. The abyssal plain is a vast area that covers almost half of Earth's surface. It is the flattest place on Earth.

But the plain is not flat everywhere. In fact, the world's highest mountains and deepest canyons are located here, beneath the ocean. The movement of Earth's crust has created the deepest parts of the ocean. Some areas plunge more than 10,000 m (33,000 ft) below the surface of the ocean. The Mariana Trench, in the Western Pacific near Japan, is 11,003 m (36,198 ft) deep. This is the deepest part of Earth's oceans.

Fast Fact

The movement of the Earth's crust can cause earthquakes on the ocean floor, producing gigantic waves called tsunamis. In 2004, movement of the Burma plate and India plate in the waters off Indonesia caused tsunami waves 30 feet high.

MAIN IDEA AND DETAILS What are the three regions of the ocean floor and what are the characteristics of each?

Oceans – Climate and Resources

Oceans affect Earth's climate in important ways. Without oceans, Earth's climate would be too harsh to support life!

Differences in water and land temperatures cause sea breezes to blow in and cool the land during the day. At night, the land cools down and land breezes blow out to sea.

Water takes much longer to heat up than land does, so oceans keep land cooler during the summer. Since water takes longer to cool down than land does, oceans keep land warmer during the winter.

Ocean water flows in steady, streamlike movements called **currents**. Currents greatly affect Earth's climate by bringing warm or cold water near land. The Gulf Stream is a current that carries warm water from the Caribbean Sea to Europe and helps keep its winters from getting too cold. The Humboldt Current carries cool water from near Antarctica to cool the west coast of South America.

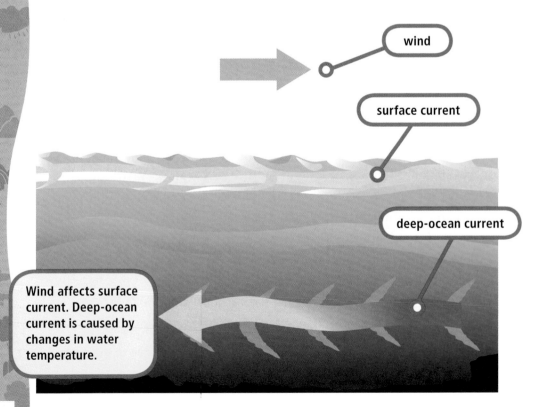

wind

surface current

deep-ocean current

Wind affects surface current. Deep-ocean current is caused by changes in water temperature.

At this processing plant in France, salt is collected from ocean water.

Oceans provide resources to support life on Earth in many important ways. Saltwater organisms—fish, shrimp, seaweed, crab—are foods which many animals rely on.

The oceans also provide us with sea salt, an ingredient for cooking. Salt is a key nutrient for farm animals, and is also used for preserving meat and refining metals.

Nonfood products also come from the ocean. Sand, used to make concrete and certain minerals, can be "mined" from the ocean and the ocean floor. Oil is pumped up from beneath the ocean floor by offshore rigs. Even jewelry—pearls from oysters—comes from the ocean.

People do not drink salt water. But in areas where freshwater is scarce, special desalinization plants remove the salt from salt water so people can use it.

MAIN IDEA AND DETAILS **How is the ocean important even for organisms that don't live in it?**

Intertidal Ecosystems

The ocean floor plunges from very shallow to very great depths. Depending on depth and location, water can be cold or warm. There is a great diversity of marine ecosystems, the many ocean environments being classified by depth and temperature.

Each ecosystem in the ocean belongs to a major ocean zone. Ocean zones are determined by water depth.

In an intertidal zone, changes in water depth mean creatures can be either underwater or directly exposed to sunlight. They can also be pounded by rough water and waves.

As ocean depth increases, there is less light. The zone which is the shallowest and gets the most sunlight is the intertidal zone. This is the area of ocean between the high-tide point and low-tide point.

Even though it receives the most light, this can be a challenging place to live. The environment is always changing as the tide goes in and out. At low tide, organisms might have to find shelter from the hot sun. At high tide, they have to survive rough water and incoming waves.

When the tide comes in and water is deep, moon jellyfish capture zooplankton with their tentacles. As the tide goes out they may be stranded on the beach.

The organisms in intertidal zones handle changes in their environment in different ways. Many hide from the sun during low tide. The sand dollar buries itself in the mud when the tide is low and there is little water to keep it moist and protected. Crabs hide under rocks.

Mussels and barnacles have shells that open and close. They close their shells during low tide, trapping water inside. During high tide, they open up and feed.

Mussels, barnacles, and many other ocean organisms depend on plankton for food. Plankton are microscopic open-water protists, animals, and bacteria that float on the ocean's surface. Plankton are important members of many marine food chains.

Most plankton use sunlight to produce food through photosynthesis. Like land organisms, most marine animals get their energy, directly or indirectly, from sunlight.

 COMPARE AND CONTRAST How does a sand dollar's way of protecting itself differ from a crab's?

Near-Shore and Open-Ocean Zones

Farther away from shore past the intertidal zone is the near-shore zone. This zone includes most of the ocean waters over the continental shelf.

The water here is relatively shallow. It gets no deeper than about 200 m (650 ft). Even the deepest water here receives sunlight. As in many aquatic ecosystems, organisms rely on producers, such as plants and algae, which are plentiful in the near-shore zone. This is a much more stable ecosystem than the intertidal zone, and is teeming with life and substantial food chains.

Many different kinds of marine life live here, from the very large to the very small. Dolphins, porpoises, sharks, and whales can be found in the near-shore zone. Plankton, shrimp, jellyfish, seaweed, and krill can also be found here.

Fast Fact

The Blue Whale, which lives in the open-ocean zone, is the world's largest animal. It can weigh 120 tons, yet feeds almost entirely on krill.

The near-shore zone is full of plant and animal life.

Extreme water pressure in deep waters crushes even heavily reinforced metal.

Past the near-shore zone is the open-ocean zone. The open-ocean zone includes most of the ocean waters above the continental slope and abyssal plain.

Waters are fairly deep in the open-ocean zone. Most of the food is found near the surface. Sunlight helps plankton grow, which in turn feeds larger consumers. Many organisms in the open-ocean zone are good swimmers, helping them search out their prey. Dolphins, seals, tuna, krill, and whales can be found in this zone.

Below about 1,000 m (3,200 ft), there is little or no sunlight. The extreme darkness means that producer organisms are hard to find.

Life in the deepest parts of this zone is also difficult because of extreme pressure from the water above. Very little life is found here. This deepest region of the open-ocean zone, which makes up about 90 percent of all the ocean waters, is cold, deep, and dark.

 COMPARE AND CONTRAST What is the difference between the deepest part of the near-shore zone and the deepest part of the open-ocean zone?

Coral Reefs

Small animals called corals form coral reefs. Corals use minerals dissolved in ocean water to form hard outer skeletons. Then the living corals attach themselves to the skeletons of dead corals. In this way, over time and very slowly, a coral reef forms.

Coral reefs are some of the largest structures on Earth that are built by living organisms. They can be quite beautiful. Coral reefs have a variety of shapes and colors. Even though they are very large, coral reefs are delicate. They can be easily damaged, and many governments set up regulations for conservation of reefs.

Living corals use minerals dissolved in ocean water to build the skeleton of coral reefs.

Reactions between the coral and the many different algae that live on the reefs are what give these structures their beautiful color. Besides the algae, a great variety of plants and animals live around reefs. In fact, even though they take up less than one percent of the ocean floor, about 25 percent of all marine organisms live in or around coral reefs.

Extensive food webs exist within coral reefs. Producers include seaweed and plankton. Many of the animals around the reef rely on plankton for food. Some types of fish eat the coral itself. Butterfly fish eat only living corals while parrotfish actually bite off parts of the reef and grind up coral skeletons. Eventually larger predators in the reef eat these fish.

 COMPARE AND CONTRAST How are butterfly fish like parrotfish? How are they different?

Coral reefs provide surfaces not far below water which sunlight can reach easily. This provides a rich home for many plants and animals in the ocean.

Deep-Ocean Vents

Most scientists have thought, until recently, that sunlight was the only energy source for Earth's ecosystems. Recently however, some researchers have found ecosystems deep in the oceans that might rely on other energy resources.

In some parts of the abyssal zone, more than 1,000 m (3,200 ft) deep, volcanic vents spew water heated to around 350°C (660°F). The water contains certain chemicals that bacteria can use, rather than sunlight, to produce and store energy.

All the other members of the ecosystem deep in the ocean rely on these bacteria. Thus, all the members of deep-ocean vents get their energy from a chemical reaction rather than from sunlight.

 COMPARE AND CONTRAST What is the difference between organisms that live in deep-ocean vents and organisms that live in other parts of the ocean?

Many of the species that rely on deep-ocean vents as their ecosystem were unknown until recently.

Summary

More than 70 percent of Earth is covered in water, but almost all of that is salt water. Very little of Earth's water is fresh water, and much of that is not usable because it is trapped in ice at the poles.

Oceans are important because they control climate. They are also important as resources of sand, oil, and other minerals. Salt water can be desalinated for use by humans.

The ocean floor can be divided into three major regions—the continental shelf, the continental slope, and the abyssal plain. The ocean itself has three major zones—intertidal, near-shore, and open-ocean. These zones and their characteristics are largely determined by depth of water.

Coral reefs provide a rich home for many ocean organisms. Deep-ocean vents are home to perhaps the only organisms on Earth that do not rely on sunlight as a source of energy.

From space it is easy to see that much of Earth's surface is covered by water.

Glossary

abyssal plain (uh•BIS•uhl PLAYN) The vast floor of the deep ocean (5, 11, 15)

continental shelf (kahnt•uhn•ENT•uhl SHELF) A gradually sloping portion of the ocean floor that is made of continental crust (4, 10, 15)

continental slope (kahnt•uhn•ENT•uhl SLOHP) The border between continental crust and oceanic crust where the ocean floor drops in depth (4, 5, 11, 15)

coral reef (KAWR•uhl REEF) A large, delicate structure built by small animals called corals (12, 13, 15)

current (KUR•uhnt) A steady, streamlike movement of ocean water (6)

evaporation (ee•vap•uh•RAY•shuhn) The process by which liquid water changes into water vapor (2)

groundwater (GROWND•waw•ter) Water located within the gaps and pores in rocks below Earth's surface (2)

precipitation (pree•sip•uh•TAY•shuhn) Solid or liquid water that falls from the air to Earth (2)

water cycle (WAW•ter SY•kuhl) The process by which water moves above, across, and through Earth's crust and ecosystems (2)